INTERNATIONAL UNION OF
PURE AND APPLIED CHEMISTRY

HOW TO NAME
AN INORGANIC SUBSTANCE
1977

A GUIDE TO THE USE OF
NOMENCLATURE OF INORGANIC CHEMISTRY:
DEFINITIVE RULES 1970

PERGAMON PRESS

HOW TO NAME AN
INORGANIC SUBSTANCE

IUPAC COMMISSION ON
THE NOMENCLATURE OF INORGANIC CHEMISTRY

The membership of the Commission during the period 1971-1975 in which the Gui
was prepared was as follows:

Titular Members

Chairman:	1971-75 W. C. Fernelius (U.S.A.)
Vice Chairman:	1971-73 K. A. Jensen (Denmark)
Secretaries:	1971-72 J. E. Prue (U.K.);
	1972-73 J. Chatt (U.K.);
	1973-75 D. M. P. Mingos (U.K.)
Members:	1971-75 R. M. Adams (U.S.A.);
	1971-73 J. Chatt (U.K.);
	1971-75 L. F. Bertello (Argentina);
	1971-75 K.-Ch. Buschbeck (Federal Republic of Germany);
	1971-75 Y. Jeannin (France);
	1973-75 G. J. Leigh (U.K.);
	1971-75 B. Myasoedov (U.S.S.R.)

Associate Members

1973-75 J. Chatt (U.K.);
1971-75 T. Erdey-Gruz (Hungary);
1973-75 K. A. Jensen (Denmark);
1971-73 G. J. Leigh (U.K.);
1971-75 W. H. Powell (U.S.A.)
1973-75 J. Riess (France);
1971-75 C. Schäffer (Denmark);
1971-75 A. A. Vlček (Czechoslovakia);
1971-73 E. Weiss (Federal Republic of Germany);
1971-75 K. Yamasaki (Japan)

International Associates

1973-75 T. D. Coyle (U.S.A.);
1973-75 J. Klikorka (Czechoslovakia);
1973-75 E. Fluck (Federal Republic of Germany).

INTERNATIONAL UNION OF
PURE AND APPLIED CHEMISTRY

INORGANIC CHEMISTRY DIVISION
COMMISSION ON NOMENCLATURE OF INORGANIC CHEMISTRY

HOW TO NAME AN
INORGANIC SUBSTANCE

being a guide to the use of
NOMENCLATURE OF INORGANIC CHEMISTRY: DEFINITIVE RULES 1970
(Second Edition published in 1971 as 'the Red Book' and in *Pure Appl. Chem.*, Vol. 28, No. 1)

also incorporating a revised and considerably enlarged section on
NAMES FOR IONS AND RADICALS
and
TABLE OF ATOMIC WEIGHTS 1975

PERGAMON PRESS

OXFORD · NEW YORK · TORONTO · SYDNEY · PARIS · FRANKFURT

U.K.	Pergamon Press Ltd., Headington Hill Hall, Oxford OX3 0BW, England
U.S.A.	Pergamon Press Inc., Maxwell House, Fairview Park, Elmsford, New York 10523, U.S.A.
CANADA	Pergamon of Canada Ltd., 75 The East Mall, Toronto, Ontario, Canada
AUSTRALIA	Pergamon Press (Aust.) Pty. Ltd., 19a Boundary Street, Rushcutters Bay, N.S.W. 2011, Australia
FRANCE	Pergamon Press SARL, 24 Rue des Ecoles, 75240 Paris, Cedex 05, France
WEST GERMANY	Pergamon Press GmbH, 6242 Kronberg-Taunus, Pferdstrasse 1, Frankfurt-am-Main, West Germany

First edition 1977

Printed in Great Britain by Page Bros (Norwich) Ltd, Norwich

ISBN 0 08 021982 9

CONTENTS

Commission on the Nomenclature of Inorganic Chemistry

HOW TO NAME AN INORGANIC SUBSTANCE

A Guide to the Use of

'Nomenclature of inorganic chemistry: definitive rules 1970'

(the Red Book)

Note of explanation. The IUPAC Commission on the Nomenclature of
Inorganic Chemistry issued first a set of rules for naming inorganic substances in 1940. Subsequent revisions were published as Definitive Rules
1957 (also called 'The Red Book') and Definitive Rules 1970 (second edition
of 'The Red Book'.[1]) Although the rules are detailed and extensive, some
chemists without a broad working knowledge of inorganic chemistry report
that they find the rules difficult to apply to specific cases. It is hoped
that this guide will aid all who have occasion to use "The Red Book" to
locate more quickly those portions of the rules (indicated by rule number)
that are pertinent to their areas of interest.

It is the nature of nomenclature that no scheme for developing names
can be utilized completely independently of nomenclature practices in other
areas of chemistry. Further, nomenclature is a continually developing
matter. Consequently, this guide contains a few references to the rules
for the 'Nomenclature of Organic Chemistry'[2,3], indicated by prefixing a
section letter (A, B, C, or D) to the rule number, and to a special report
on the 'Nomenclature of Inorganic Boron Compounds'[4]. Section D of the
organic rules[2] is particularly relevant to this guide because it is a joint
effort of the Commissions on Nomenclature of Organic Chemistry and of
Inorganic Chemistry and because it covers organometallic compounds, chains
and rings with regular patterns of heteroatoms, and organic compounds containing phosphorus, arsenic, antimony, bismuth, silicon and boron.

Introduction. There is as yet no single pattern for the naming of
inorganic substances. Instead, inorganic materials are divided into several
major groups for each of which a definite pattern of nomenclature has de-
veloped. This outline is designed to aid a person with a specific problem
in naming substances, primarily inorganic substances, to find quickly the
rules pertinent to his particular problem.

What is an inorganic compound? Inorganic substances may be defined
broadly as substances consisting of combinations of all the elements except
those that consist mainly of certain chains and rings of carbon atoms with
defined atoms and/or groups attached to these skeletal atoms. Such compounds
are treated by the principles of organic chemical nomenclature[2,3]. However,
the number of compounds known which are clearly the concern of organic

1

nomenclature presently far exceeds the number that are clearly inorganic. Moreover, the development of chemistry in this century has been such that the line of distinction between the two major branches of chemistry has been clouded, if not eliminated altogether in some cases. Derivatives of the molecular hydrides of the elements such as RPH_2, R_3As, etc. (§ D-5-7), organometallic compounds, such as R_2SnCl_2 (§ D-3), and particularly the rapidly growing area of coordination compounds (§ Sec. 7; § D-3) constitute areas that are of concern to both groups. Thus, further development of nomenclature demands cooperation between inorganic and organic chemists and an understanding by each group of the basic principles of nomenclature used by each.

Group names. Because of the similarity in behaviour of two or more elements or substances, group names are often needed to refer to small or large collections of elements or substances, e.g., halogens, transition elements (see § 1.2), berthollides (see § 9).

Background principles involved in or related to nomenclature. Oxidation number (§ 0.1); coordination number (§ 0.2); multiplying affixes (§ 0.31); enclosing marks (§ 0.33); use of italic letters (§ 0.34).

Types of names used in chemistry. Chemistry deals with a number of different types of compounds. As a result different types of names are employed. Some of them are more frequent in inorganic chemistry, some others in organic chemistry. When a compound contains simultaneously groups of atoms that are usually referred to as inorganic groups and others that are considered as organic groups, several types of name segments may occur within the same name.

1. Simple binary names: carbon disulfide, CS_2; silicon dioxide, SiO_2; phosphorus trichloride, PCl_3; copper(II) chloride, $CuCl_2$; etc. (Note that names of this type are also used for many compounds containing more than two different kinds of atoms.)

2. Molecular hydride or -ane names: germane, GeH_4; diborane, B_2H_6; etc. Many trivial names still persist in this area: ammonia, NH_3, hydrazine, N_2H_4; phosphine, PH_3; etc.

3. Substitutive names (§ C-0.1): chlorosilane, SiH_3Cl; tetramethylsilane, $(CH_3)_4Si$; dimethylphosphine, $(CH_3)_2PH$; 2-bromo-3-chloropentaborane(9), B_5H_7BrCl; etc. Here one uses names for radicals (substituents) (either inorganic or organic), parent compounds, and the principal group as well as locants.

4. Cations derived by proton addition to molecular hydrides (or their derivatives): ammonium bromide NH_4Br or $[NH_4]^+$, $Br-$; phosphonium chloride PH_4Cl or $[PH_4]^+$, Cl^-; guanidinium hydrogen sulfate $[CH_6N_3]^+$, $[HSO_4]^-$; tetramethylammonium hexafluorophosphate, $[(CH_3)_4]^+$, $[PF_6]^-$.

5. Oxo acids and their salts (use of the affixes hypo-, -ous, -ite, -ic, -ate, and per-): nitric acid, HNO_3; sodium nitrite, $NaNO_2$; periodic acid, H_5IO_6.

6. Condensed acids and their salts: diphosphoric acid $H_4P_2O_7$; sodium cyclo-triphosphate, $Na_3P_3O_9$; dodecamolybdophosphoric acid, $H_3PMo_{12}O_{40}$.

7. <u>Mixed salts</u>: magnesium chloride hydroxide MgCl(OH); potassium sodium carbonate, $KNaCO_3$.

8. <u>Replacement names</u>: thiosulfuric acid, $H_2S_2O_3$; selenocyanate ion, $SeCN^-$; potassium trithiocarbonate, K_2CS_3; dicarba-<u>closo</u>-pentaborane(5). $B_3C_2H_5$.

9. <u>Coordination names</u> (§ 7).

10. <u>Functional class names</u>: carbonic anhydride, CO_2; phosphoryl chloride, $POCl_3$; sulfuric diamide, $SO_2(NH_2)_2$; etc. (§ 5.3). Names of this type are still used but are not recommended, especially for anhydrides.

11. <u>Functional suffix nomenclature</u>: hydrazinesulfonic acid, H_2NNHSO_2OH. (§ 3.33).

12. <u>Additive names</u>: triphenylphosphine oxide $(C_6H_5)_3PO$ (§ D-5.4), ammonia-boron trifluoride (1/1), $H_3N \cdot BF_3$ (§ 8).

13. <u>Subtractive names</u>. In certain fields of organic chemistry, use is made of a type of name indicating removal of atoms or groups from a compound denoted by a systematic or trivial name: de-<u>N</u>-methylmorphine (removal of CH_2); 6-deoxy-α-D-glucopyranose (removal of $-O-$) (§ C-0.4). Inorganic nomenclature has made little use of such names although they now seem to offer the best solution to problems encountered among some of the boron clusters (§ 7.5 of ref. 4).

<u>Procedure</u>

1. Is the substance an element?

a. See § 1.1 and Table I. (A table of elements with symbols, atomic numbers and atomic weights is appended to this Guide.)

b. In addition to the simple name, it may be desirable to designate the allotropic modification (§ 1.4).

c. In case names for groups of elements are desired see § 1.2.

d. In case it is desired to refer to a specific isotope of an element see § 1.15, 1.3.

2. Is there need to designate the presence of a specific isotope in a compound? See § 1.32 and then the relevant section below.

3. Is the substance a binary compound (two different elements)?

a. See § 2.2 for general treatment (also § 3.1 - 3.3) and § 2.3 if the substance furnishes a homopolyatomic anion (N_3^-, $S_x^=$, I_3^-, etc.) (also § 3.221).

b. There are special rules for boron hydrides. See § 11 and ref. 4.

    c. If the hydride is a chain, see § 2.3 and D-4; or a ring, see § D-4.

    d. A substance of non-stoichiometric composition requires special treatment. See § 9.

    e. Is the substance a radical? See § 3.3 and Table II.

    4. Is the substance a ternary (three different elements) compound?

    a. Some ternary compounds ($NaNH_2$, KCN, $Ca(OH)_2$, etc.) are treated as pseudobinary compounds. See § 2.2, 3.13 - 3.16 and 3.221 - 3.223. If it is a radical see § 3.3 and Table II.

    b. Can the substance be considered as a derivative of a binary compound?

    i. For a substitution product of a volatile hydride ($ClNH_2$, $BrSiH_3$, etc.) see § 2.3, 3.221, D-5.1, and D-7.4. For derivatives of boron hydrides, see § 11 and ref. 4.

    ii. For a substitution product of a homogeneous chain or ring see § D-4.1, D-4.2, and D-6.1.

    iii. For a replacement or substitution product of a borane see § 11.2, 11.4, 11.7, D-7.2, D-7.4, and elsewhere[4].

    c. For oxo acids, see § 5.2; for their salts, see § 3.223 and 3.224 and 6. For condensed oxo acids, see "isopoly acids" below.

    d. For double oxides, see § 6.5.

    e. For thio acids (K_2CS_3, etc.), see § 5.23.

    f. For chloro acids (H_2PtCl_6, etc.), see § 5.24.

    g. For chain and ring isopoly acids (polyphosphates, etc.), see § 4.11 - 4.14.

    h. For chains and rings of repeating units see §D-4.4, D-4.5, and D-6.2 - 6.4.

    i. For non-stoichiometric compounds, see § 9.

    5. Is the substance a quaternary (four different elements) or higher order compound, exclusive of a coordination, cluster, or addition compound?

    a. Can the substance be considered as a derivative of a binary or ternary compound? See appropriate entries under item 3. For oxo acids, see § 5.2; for other acids, see § 5.23, 5.24 and 5.3.

    b. Is the substance a heteropoly acid ($H_3PW_{12}O_{40}$, etc.)? See § 4.2.

    c. Is the substance an acid, basic, or double salt? See § 6.2 - 6.4.

d. Is the substance a double oxide or hydroxide? See § 6.5

e. Is the substance a substitution or replacement derivative of a borane? See § 11.2, 11.4, and 11.7.

6. Is the substance a coordination compound $(K_3[Co(C_2O_4)_3]$ $[Pt(NH_3)_6]Cl_4$, etc.)? See § 7.1 – 7.6 and 7.71.

7. Is the substance a cluster compound $([Mo_6Cl_6](SO_4)_2$, $Rh_6(CO)_{16}$, etc.)? See § 7.72.

8. Is the substance an addition compound $(H_3N \cdot BF_3$, etc.)? See § 8.

References

1. Nomenclature of Inorganic Chemistry, Definitive Rules 1970, (2nd Edn.) Butterworths, London, 1971, 110 pp. Also published in Pure Appl. Chem. 28, 1 (1971).

2. Nomenclature of Organic Chemistry, Sections A, B, and C, (3rd Edn.) Butterworths, London, 1971, 352 pp.

3. Nomenclature of Organic Chemistry, Section D, Tentative Nomenclature Appendix No. 31 (Aug. '73) IUPAC Information Bull. No. 31, Aug. 1973. IUPAC Secretariat, Oxford, U.K.

4. Nomenclature of Inorganic Boron Compounds, Pure Appl. Chem. 30 (3-4), 683 (1972).

5. Combined Introductions, Indexes to Volume 66; Chem. Abstr. 66, 281-401 (1967).

6. Selection of Index Names for Chemical Substances, Chem. Abstr., 76, 1261-1331 (1972).

NAMES FOR IONS AND RADICALS

Names for ions and radicals were given in Table II of <u>Nomenclature of</u> <u>Inorganic</u> <u>Chemistry</u>[1]. The revised Table II which follows is intended to be a convenient reference guide to approved names or name fragments. The formulas are listed alphabetically by atoms and then according to the number of each in turn in the formula rather than by total number of atoms of each element. Whenever the empirical formula so written gives little or no clue as to the structure of the unit, the formula is rewritten in the normal manner. Along with the items in this table are given references to this and other compilations of acceptable nomenclature. A number without any letter symbol refers to the Inorganic Rules[1]; the symbol $\sim$ preceding a number indicates that, although the item is not mentioned specifically in a rule, it is completely in harmony with an example which is mentioned specifically in the Inortanic Rules. A number preceded by Bo refers to the Rules for the Nomenclature of Inorganic Boron Compounds[4]. A number preceded by an A refers to Section A of the Organic Rules[2]; a number preceded by a C refers to Section C of the Organic Rules[3]; and a number preceded by a D refers to Section D of the Organic Rules[3]. CA 66 refers to Tables of Groups and Radicals in the Introduction to the Indexes to Volume 66 of <u>Chemical</u> <u>Abstracts</u>[5] and CA 76 to similar tables in the Index Guide to Volume 76 of <u>Chemical</u> <u>Abstracts</u>[6]. Entries in the original table are underlined.

The table contains six columns: 1) the empirical formula, 2) the name for an uncharged atom, molecule or radical of that composition, 3) the name for a cation or cationic radical of that composition, 4) the name of the anion of that composition, 5) the name of the ligand of that composition and 6) the prefix for the group of that composition as used in substitutive nomenclature. The entries in the sixth column are used much more by organic chemists than by inorganic chemists. They are included here for two reasons (1) for comparison with inorganic practices and (2) as a guide for inorganic chemists when dealing with organic ligands, substitution products of molecular hydrides, addition compounds, etc. In many instances throughout the table there are blank spaces in the table. This does not imply that names for these entities are not justified on the basis of existing rules but that there is seldom, if ever, a need for such names especially by inorganic chemists.

There are a few entries in this table which are not covered by specific IUPAC rules. This means that the Commission as yet has made no recommendation covering them. In such cases, if Chemical Abstracts has a definite policy covering such a case, the name with the authority is listed. The inclusion is solely for the convenience of the user and does not imply the Commission's endorsement of the practice.

For many compounds, the IUPAC rules require that the name contains either the oxidation number (Roman numerals; often called the Stock number)

(§ 0.1) or the charge on an ion (Arabic numeral; often called the Ewens-Bassett number). In this table, only the charge on the ion is given. The choice is one of convenience and should not be interpreted either as a recommendation or as a preference by the Commission.

Atom or group	NAME				
	as uncharged atom, molecule or radical	as cation or cationic radical	as anion	as ligand	as prefix in substitutive nomenclature
As	(mono)arsenic 1.4	arsenic(1+) ~3.11	arsenide(3-) 3.21	arsenido 7.311	arsinetriyl, As≡ D-5.15; arsinidyne, As≡ CA 76; arsylidyne, As≡ CA 66
AsH	(mono)hydrogen (mono)arsenic	hydrogenarsenic(1+)	hydrogenarsenide ~3.22	hydrogenarsenido ~7.311	arsinediyl, HAs= D-5.14; arsinidine, HAs= CA 76; arsylene, HAs= CA 66
AsH$_2$	dihydrogen arsenic	dihydrogenarsenic(1+)	dihydrogenarsenide ~3.222	dihydrogenarsenido ~7.311	arsino, H$_2$As- D-5.12; CA 76; arsoranylidyne, H2As≡ CA 76
AsH$_3$	arsine 2.3; D-5.11; arsane 2.3	trihydrogenarsenic(1+)	. . .	arsine 7.321	arsonio, H$_3^+$As- D-5.13; arsoranediyl, H$_3$As= ~D-5.74; arsoranylidene, H$_3$As= CA 76
AsH$_4$	. . .	arsonium 3.14; D-5.31	. . .	. . .	arsoranyl, H$_4$As- D-5.74; CA 76
AsH$_5$	arsorane D-5.71; λ^5-arsane D-0.3	. . .	. . .	. . .	. . .
AsO$_3$	. . .	. . .	arsenite 3.224	arsenito 7.311	. . .
AsO$_4$	. . .	. . .	orthoarsenate; arsenate 5.214; tetraoxoarsenate(V) 5.214	arsenato 7.311	. . .
As$_2$	diarsenic 1.4	. . .	. . .	. . .	arseno*, -As=As- CA 66; 1,2-diarsenediyl*, -As=As- CA 76; diarsinetetrayl, =As-As- CA 66; 1,2-diarsinediylidene CA 76
As$_2$H, HAs=As-	. . .	. . .	. . .	. . .	diarsenyl, HAs=As- CA 76
As$_2$H$_3$	. . .	. . .	. . .	. . .	diarsinyl, H$_2$As-AsH- CA 76; diarsanyl 2.3; D-4.14
As$_2$H$_4$	diarsane 2.3	. . .	. . .	. . .	. . .

* The names arseno and 1,2-diarsenediyl have been used for the group -As=As- and diarsenyl for HAs=As- even though the substances once thought to contain these groups are now known to be polymeric. Hence, they are needed only as class names or in describing hypothetical compounds.

Formula					
B	(mono)boron 1.1	boron(+) 3.1	boride 2.22 3.21	borido ~7.312	boranetriyl, B≡ Bo-4.1; D-7.3 / borylidyne, B≡ CA 76
$B_xC_yH_z$	see Bo-7; D-7.2 for individual compounds	···	see Bo-7.5 for specific examples	see Bo-7.2 & 7.3 for specific examples	see Bo-7.4; CA 76, §161 for specific examples
BH	borane(1) 11.11	hydroboron(1+) Bo-5.3	hydroborate(2-) Bo-5.3	hydroborato	boranediyl, HB= Bo-4.1; D-7.3 / hydroboric, HB= Bo-4.1 / borylene, BH= CA 76
BHO_2·$(HBO_2)_n$	metaboric acid	···	···	···	···
BH_2	borane(2) 11.11	dihydroboron(1+) Bo-5.3	dihydroborate(1-) Bo-5.3	dihydroborato	boryl, H_2B- Bo-4.1; D-7.3; CA 76 / dihydroborio Bo-4.1
B_2H_2	diborane(2) 11.11	dihydrodiboron(1+)	dihydrodiborate(1-)	dihydrodiborato(1-)	1,2-diborane(4)diyl, -HB-BH- Bo-4.21 / 1,1-diborane(4)diyl, H_2B-B= Bo-4.21
BH_2O_2, $(HO)_2B$-	···	···	···	···	dihydroxyboryl, $(HO)_2B$- Bo-4.1 / borono, $(HO)_2B$- CA 76 / dihydroxyborio, $(HO)_2B$- Bo-4.1
BH_3O_3 H_3BO_3	orthoboric acid boric acid	···	···	···	···
BO	···	···	···	···	oxoboryl, OB- Bo-4.1; D-4.1; CA 76 / oxoborio, OB- Bo-4.1; D-4.1
BO_2	···	···	metaborate 3.223; 5.214	···	···
BO_3	···	···	orthoborate 3.223; 5.214	···	···
B_2H_3	diborane(3) 11.11	trihydrodiboron(1+) Bo-5.3	trihydrodiborate(1-) Bo-5.3	···	diborane(4)yl, H_2B-BH- Bo-4.21
B_2H_4	diborane(4) 11.11	tetrahydrodiboron(1+) Bo-5.3	tetrahydrodiborate(1-) Bo-5.3	···	1,1-diborane(6)diyl, $H_2B(H)_2B$= Bo-4.2 / 1,2-diborane(6)diyl, -$HB(H)_2BH$- Bo-4.2
B_2H_5	diborane(5) 11.11	pentahydrodiboron(1+) Bo-5.3	Pentahydrodiborate(1-) Bo-5.3	···	diborane(6)yl, $H_2B(H)_2BH$- Bo-4.21

Atom or group	NAME				
	as uncharged atom, molecule or radical	as cation or cationic radical	as anion	as ligand	as prefix in substitutive nomenclature
B_4O_7	...	...	tetraborate 4.12	...	...
B_xH_y.....	see Bo-2 & D-7.1 for individual compounds	see Bo-5 & D-7.6 for specific examples	see Bo-5 & D-7.6 for specific examples	...	see Bo-4.2; CA 76, §161i & D-7.3 for specific examples
Bi	(mono)bismuth 1.4	bismuth(3+) 3.11	bismuthide 2.22; ~3.21	bismuthido 7.311	bismuthinetriyl, Bi≡ D-5.15; bismuthylidyne, Bi≡ CA 76
BiH	...	...	hydrogenbismuthide 3.22	...	bismuthinediyl, HBi= D-5.14; bismuthylene, HBi= CA 76
BiH_2	...	...	dihydrogenbismuthide 3.22		bismuthino, H₂Bi- D-5.12; CA 76
BiH_3	bismuthine 2.3; bismuthane 2.3				bismuthonio, H₃Bi⁺- C-5.33
Bi_9	...	nonabismuth(5+) ~3.11	...	...	...
Br	(mono)bromine 1.4	bromine(+) ~3.11	bromide 3.21	bromo 7.312; CA 76	bromo, Br- C-10.1; C-102; CA 76
BrF_2	(mono)bromine difluoride 2.251	difluorobromine(1+) 3.13	difluorobromate(1-) ~5.24; 3.223	difluorobromato(1-) 3.223; 7.311	difluorobromo, F₂Br- ~C-106.3
BrH HBr	hydrogen bromide 2.21, 2.22	hydrogenbromine(1+) 3.18; bromoniumyl H⁺Br- C-83.3	...	...	bromonio, H⁺Br- C-82.1
BrH_2 H_2Br	...	bromonium H₂Br⁺ C-82.1			...
BrO	(mono)bromine (mono)-oxide 2.21; 2.22; 2.251	bromosyl cation 3.12; 3.32; oxobromine(1+) 3.13	hypobromite 3.224; 5.214; (mono)oxobromate(1-) 2.24; 3.223	hypobromito 7.311; oxobromato(1-) 7.311	bromosyl C-106; 3.12; 3.32
BrO_2	(mono)bromine dioxide 2.21; 2.22; 2.251	bromyl cation ~3.12; dioxobromine 3.32; 3.13	bromite 5.214; dioxobromate(1-) 2.24; 3.223	bromito 2.3111; dioxobromato(1-) 7.311	bromyl C-10.1; C-106.2
BrO_3	(mono)bromine trioxide 2.21; 2.22	trioxobromine(1+) ~3.13; perbromyl cation ~3.12	bromate 5.214; trioxobromate(1-) 2.24; 3.223	bromato 7.311; trioxobromato(1-) 7.311	perbromyl, O3Br- ~C-10.1; C-106.6
BrO_4	(mono)bromine tetraoxide 2.21; 2.22; 2.251	tetraoxobromine(1+) 3.13	perbromate 5.214; tetraoxobromate(1-) 2.24; 3.223	perbromato 7.311; tetraoxobromato(1-)	...

C	(mono)carbon 1.4	· · ·	carbide 2.22; 3.21	carbido 7.311 methanetetrayl 7.313	methanetetrayl, C≡ ∿D-4.4; CA 76 carbyl, -C- CA 76
CH	· · ·	formyl cation C-83.1 formylium C-83.1	· · ·	methylidyne A-4.1; 7.313	methylidyne, HC≡ A-4.1; CA 76
CHO HC(O)	formyl C-81.1; C-404.1	· · ·	· · ·	· · ·	formyl, HC(O)- C-10.3; C-304.2; C-404.1; CA 76
CHOS HS(O)C HO(S)C	thiocarboxyl C-81.1 C-541.1	· · ·	HC(O)S thioformate 5.23	thioformato 7.311	thiocarboxy, HS(O)C- C-541.1; CA 76
CHO2 HOC(O)	carboxyl C-81.1; C-401.3	· · ·	formate C-404.1	formato, HC(O)O 7.311	carboxy, -COOH C-10.3; C-401.2; CA 76 formyloxy, HC(O)O- C-463.3
CHO3 HCO3	· · ·	· · ·	hydrogencarbonate 3.222	hydrogencarbonato 3.222; 7.322	carboxyoxy, HOC(O)O- ∿CA 76 formyldioxy HC(O)OO- ∿C-463.3
CHS SCH	thioformyl C-81.1; C-531.3; C-543.4	thioformyl cation C-83.1 thioformylium C-83.1	· · ·	· · ·	thioformyl, SCH- C-531.3; C-543.4; CA 66 thioxomethyl, SCH- CA 76
CH2	methylene A-4.1; C-81.1 carbene	methylene dication C-83.1 methanediylium C-83.1	methanediide ∿C-84.3	methylene, CH = A-4.1; 7.313	methylene, CH2= A-4.1; CA 76
CH2N HC(NH)	formimidoyl C-81.1 C-451.2	· · ·	· · ·	· · ·	formimidoyl, HC(NH)- C-451.2; CA 66 iminomethyl, HC(NH)- CA 76
CH2N H2C=N	methyleneaminyl C-81.2	methyleneaminyl cation C-83.1 methyleneaminylium C-83.1	H2C=N- methaniminate 7.314; C-815.3 methyleneiminate methyleneamide	methaniminato methyleneaminato methyleneamido	methyleneimino, H2C=N- C-815.3
CH2NO NH2CO	carbamoyl C-81.1; C-431.2	carbamoylium C-83.1 carbamoyl cation C-83.1	· · ·	carbamoyl C-431.2	aminocarbonyl, H2NC(O)- CA 76 carbamoyl C-431.2; CA 66 formamido HC(O)NH- C-823.1; CA 66 formylamino CA 76
CH2NO2 NH2CO2	· · ·	· · ·	carbamate C-431.1	carbamato 7.311; C-431.1	carbamoyloxy, NH2C(O)-
CH2NO2 O2NCH2	· · ·	· · ·	nitromethanate, O2NCH2⁻	· · ·	nitromethyl, O2NCH2-

		NAME			
Atom or group	as uncharged atom, molecule or radical	as cation or cationic radical	as anion	as ligand	as prefix in substitutive nomenclature
CH₂NS H₂NC(S)-	thiocarbamoyl C-547.1	...	...	...	thiocarbamoyl, H₂NC(S) CA 76 aminothioxomethyl C-547.1; CA 76
CH₂N₂O H₂NC(O)N	carbamoylaminylene C-81.2	...	...	...	carbamoylimino, H₂NC(O)N= C-815.3; CA 66 (aminocarbonyl)imino CA 76
CH₂N₂O HNC(O)NH	carbonyldiaminyl C-81.2	...	...	carbonyldiamido 7.311 ureato(2-) 7.314	ureylene, -HNC(O)NH- CA 66 C-72.2; C-971.3 carbonyldiimino CA 76
CH₃	methyl A-1.2; C-81.1	carbenium C-83.1 methylium C-83.1 methyl cation C-83.1; 3.12	methanide C-84.3	methyl A-1.2; 7.313	methyl, H₃C- A-1.2; CA 76
CH₃O	methoxyl C-81.1; C-205.1	methoxyl cation C-83.1 methoxylium C-83.1 methyloxygen(1+) 3.13	methoxide C-206.2 methanolate C-206.1 methyl oxide C-206	methoxo 7.312 methanolato 7.312 methoxy 7.312 fn.	methoxy, CH₃O- A-1.2; C-205.1; CA 76
CH₃N₂ H₂NC(=NH)	...	...	...	...	amidino, H₂NC(=NH)-* C-951.4; CA 66 aminoiminomethyl CA 76
CH₃N₂ HC(=NH)NH	...	...	...	...	iminomethylamino, C-951.4; CA 76 formimidoylamino CA 66
CH₃N₂O	...	...	...	...	ureido, H₂NCONH- C-971.2; CA 66 [(aminocarbonyl)amino] CA 76
CH₃N₂O	...	...	...	...	carbazoyl, H₂NNHCO- C-984.1; CA 66 hydrazinocarbonyl CA 76
CH₃S	methylsulfanyl C-81.1 methylthio C-81.1 methylsulfenyl C-81.1	methylsulfur(1+) ~3.13 methylsulfanyl cation C-83.1 methanesulfenyl cation C-83.1 methylsulfanylium C-83.1	methanethiolate 7.312 methylsulfide ~C-206	methanethiolato 7.312 methylthio 7.312	methylthio, CH₃S- C-514.1; CA 76

*The name guanyl is no longer recommended either by IUPAC or Ca.

CH₄	methane A-1.1	methanesulfenylium C-83.1 methane cation C-83.3 methaniumyl C-83.3	· · ·	methane	methaniumyl, H₄C⁺ A-1.2; C-83.3
CH₄N CH₃NH-	methylaminyl C-81.2	methylaminylium C-81.2; C-83.1 methylaminyl cation	methylamide 3.221; ~7.311 methylaminate 7.314; ~C-87	methylamido ~7.311 methylaminato 7.314	methylamino, CH₃NH- C-811.4 methylimino, CH₃NH= C-82.1; C-82.2
CH₄N₃	· · ·	· · ·	· · ·	· · ·	guanidino, H₂NC(=NH)NH- C-961.2; CA 66 (aminoiminomethyl)amino, CA 76 aminoamidino, H₂NHHC(=NH)- CA 66 hydrazinoiminomethyl CA 76 aminohydrazonomethyl, H₂N-NC(NH₂)- CA 76
CH₅		methanium H₅C⁺			
CC1O ClC(O)	chloroformyl, C-10.3 C-81.1; C-481.2	chloroformylium, C-83.1 chloroformylium cation C-83.1 chlorooxomethylium ~C-83.1	· · ·	chloroformyl 7.313 chlorocarbonyl	chloroformyl, ClC(O)- C-10.3; C-481.2 CA 66 chlorocarbonyl, ClC(O)- CA 76
CN	(mono)cyanogen ~1.4	cyanogen	cyanide 3.221	cyano 7.312 cyano-C ~7.33	cyano, NC- C-10.2; C-832.5; CA 76
CN, NC	· · ·	· · ·	· · ·	isocyano ~7.33; CA 66 cyano-N ~7.33	isocyano, CN- C-833.1; CA 76
CN₂	· · ·	· · ·	cyanamidate(2-)	carbodiimidato(2-) 7.314 methanetetrayldiamido 7.311 cyanamidato(2-) 7.314	methanetetraylnitrilo, -N=C=N- CA 66 cyanimino, N≡C-N=
CNO, OCN	· · ·	· · ·	cyanate 5.214	cyanato ~7.311; 7.33 cyanato-O 7.33	cyanato, NCO- C-833.1; CA 76
CNO, NCO	· · ·	· · ·	cyanate 5.214	isocyanato ~7.311; 7.33 cyanato-N 7.33	isocyanato, OCN- C-833.1; CA 76
CNO, ONC	· · ·	· · ·	fulminate 5.214	fulminato ~7.311	· · ·
CNS, SCN	thiocyanogen	thiocyanogen	thiocyanate 5.23	thiocyanato ~7.311; 7.33 thiocyanato-S 7.33	thiocyanato, NCS- C-833.1; CA 76

13

NAME

Atom or group	as uncharged atom, molecule or radical	as cation or cationic radical	an anion	as ligand	as prefix in substitutive nomenclature
CNS, NCS	. . .	. . .	thiocyanate	isothiocyanato ~7.311; 7.33; thiocyanato-N 7.33	isothiocyanato, SCN- C-833.1; CA 76
CNSe SeCN	selenocyanogen	selenocyanogen	selenocyanate ~5.214	selenocyanato ~7.33; selenocyanato-Se ~7.33	selenocyanato, NCSe- C-833.1; CA 76
CNSe NCSe	. . .	. . .	selenocyanate	isoselenocyanato ~7.33; selenocyanato-N ~7.33	isoselenocyanato, SeCN- C-833.1
CO	carbon monooxide 2.22; 2.24; 2.251	carbonyl 3.32	. . .	carbonyl 7.323	carbonyl, OC= C-72.1; C-108.2; C-403.2; CA 76
CO$_2$	carbon dioxide 2.22; 2.24	. . .	. . .	carbon dioxide 7.321	carboxylato, -OC(O)- C-86.1
CO$_3$	. . .	. . .	carbonate 5.214	carbonato 7.311	carbonyldioxy, -O-C(O)-O- C-205.2 CA 66; carbonylbis(oxy) CA 76
CS	carbon monosulfide 2.22; 2.24	thiocarbonyl ~3.32	. . .	thiocarbonyl ~7.323	thiocarbonyl, -SC- C-108.2; C-543.2; C-545.1; CA 66; carbonothioyl, SC= CA 76
CS$_2$	carbon disulfide 2.21; 2.22; 2.251	. . .	. . .	carbon disulfide 7.321	dithiocarboxylate, -S-C(S)- C-86.1
C$_2$	dicarbon 1.4 ethynylene A-4.3; C-81.1	. . .	acetylide 3.211; C-84.3 ethynediide ~C-84.3	ethynylene 7.313 ethynediyl ~7.313	ethynylene, -C≡C- CA 66; A-4.3 ethynediyl, -C≡C ~CA 76
C$_2$H	ethynyl A-3.5; C-81.1	ethynylium C-83.1	ethynide C-84.3	ethynyl 7.313	ethynyl, HC≡C- A-3.5
C$_2$H$_2$	acetylene A-3.2	ethenediylium ~C-83.1	ethenediide, ~C-84.3	acetylene 7.321 vinylene 7.313	vinylene, -CH=CH- A-4.3; CA 66 1,2-ethenediyl, -CH=CH- CA 76 vinylidene, H$_2$C=C= A-4.1; CA 66 ethenylidene, H$_2$C=C= CA 76
C$_2$H$_2$NO$_2$ H$_2$NCOCO-	oxamoyl C-81.1; C-431.2	. . .	. . .	. . .	oxamoyl, H$_2$NCOCO- C-431.2 CA 66 aminooxoacetyl CA 76

14

Formula					
$N{\equiv}C\text{-}COOH$	cyanoacetic acid	· · ·	· · ·	· · ·	· · ·
C_2H_2O $CH_2{=}C{=}O$	ketene $CH_2{=}C{=}O$	· · ·	· · ·	· · ·	carbonylmethylene-, $-COCH_2-$ CA 66; 1-oxo-1,2-ethanediyl CA 76; oxoethylene, $-COCH_2-$ CA 66
C_2H_3 $H_2C{=}CH$	vinyl A-3.5	vinylium A-3.5; C-83.1	ethenide C-84.3	vinyl 7.313	vinyl, $H_2C{=}CH-$ A-3.5 CA 66; ethnyl, $H_2C{=}CH-$ CA 76
C_2H_3NO $CH_3C(O)N$	· · ·	· · ·	acetylimide 3.221	· · ·	acetylimino, $CH_3(O)N{=}$ C-815.3
C_2H_3O CH_3CO	acetyl C-81.1; C-404.1	acetyl cation $\sim$3.12 C-83.1; acetylium C-83.1	· · ·	acetyl $\sim$7.313	acetyl, $CH_3C(O)O-$ C-404.1; CA 76
$C_2H_3O_2$ $CH_3C(O)O$	acetoxyl C-81.i; C-463.3	acetoxyl cation C-83.1; acetoxylium C-83.1	acetate C-84.1	acetato 7.311	acetoxy, $CH_3(O)-$ C-463.3; CA 66; acetyloxy CA 76
C_2H_3S $CH_3C(S)$	thioacetyl C-81.1; C-541.2	thioacetyl cation C-83.1; thioacetylium C-83.1	· · ·	· · ·	thioacetyl, $CH_3C(S)-$ C-541.2
C_2H_4	ethylene A-3.1; ethene A-3.1	ethanediylium C-83.1; ethylene dication	ethanediide, C-84.3; ethene anion C-84.4	ethylene 7.314; 7.421; ethene A-3.1; 7.314	ethylene, $-CH_2{=}CH_2-$ A-4.2; CA 66; 1,2-ethanediyl, $-CH_2\text{-}CH_2-$ CA 76; ethylidene, $CH_3CH{=}$ A-4.1; CA 76
C_2H_4N $CH_3C({=}NH)-$	· · ·	· · ·	· · ·	· · ·	acetimidoyl, $CH_3C({=}NH)-$ C-451.2; CA 66; 1-iminoethyl CA 76
C_2H_4NO $CH_3C(O)NH-$	acetylaminyl C-81.2	acetylaminylium C-81.2; C-83.1	acetylamide 3.221; acetamidate $\sim$C-87	acetamido 7.311; acetamidato 7.311; 7.314	acetamido, $CH_3C(O)NH-$ C-823.1 CA 66; acetylamino CA 76
$C_2H_4O_2$ OCH_2CH_2O	ethylenedioxyl C-81.1; C-250.2	ethylenedioxylium C-83.1	1,2-ethanediolate C-84.2; C-206; ethylene glycolate C-84.2; C-206	1,2-ethanediolato(2-) 7.314; ethylene glycolato(2-)	ethylenedioxy, $-OCH_2CH_2O-$ C-72.2; C-205.2; C-212.3; 1,2-ethanediylbis(oxy) CA 76
C_2H_5O	ethoxyl C-81.1	ethoxyl cation C-83.1; ethoxylium	ethanolate, ethoxide C-206.1; ethyl oxide	ethanolato 7.312; ethoxo	ethoxy, CH_3CH_2O- C-205.1

Atom or group	NAME				
	as uncharged atom, molecule or radical	as cation or cationic radical	an anion	as ligand	as prefix in substitutive nomenclature
C_2O_2	...	...	...	...	oxalyl, $-C(O)-C(O)-$ C-404.1; C-405.2 CA 66; 1,2-dioxo-1,2-ethanediyl CA 76
C_2O_4	...	...	oxalate, C-401.1; C-461.1; ethanedioate, C-461.1	oxalato 7.311; ethanedioato 7.311	...
C_3H_2	...	...	...	...	2-propynylidene, $HC{\equiv}CCH=$ A-3.5; A-4.1; CA 66
C_3H_3	propynyl A-3.51	propynylium C-83.1	propynide C-84.3	propynyl 7.314	1-propynyl, $H_3CC{\equiv}C-$ A-3.51 CA 66; 2-propynyl, $HC{\equiv}C-CH_2-$ A-3.51
C_3H_3O	...	...	...	...	acryloyl, $CH_2=CHCO-$ C-404.1; CA 66; 1-oxo-2-propenyl CA 76; 1-carbonylethyl; $O=C=CMe-$ CA 66; methyloxoethenyl CA 76
C_3H_4	...	...	...	...	allylidene, $CH_2=CH-CH=$ A-3.5; A-4.1; CA 66; 2-propenylidene, A-4.3; CA 76; 1-propenylidene, $CH_3CH=C=$ A-4.1; CA 76
C_3H_5	...	...	...	...	allyl, $CH_2=CHCH_2-$ A-3.5; CA 66; 2-propenyl CA 76; 1-propenyl, $CH_3-CH=CH-$ CA 76; cyclopropyl, A-11.2
C_3H_6	...	...	...	...	isopropylidene, $Me_2C=$ A-4.1; CA 66; 1-methylethylidene, CA 76; propylidene, $CH_3CH_2CH=$ A-4.1; CA 76

Continuation (preceding entry):

propylene, −CHMeCH₂−, A−4.2; CA 66; 1-methyl-1,2-ethanediyl, CA 76; trimethylene, −CH₂CH₂CH₂− A−4.2; CA 66; 1,3-propanediyl, CA 76

Formula					
C_3O_2 O=C=C=C=O	tricarbon dioxide 2.251	· · ·	· · ·	· · ·	· · ·
C_4H_4	· · ·	· · ·	· · ·	· · ·	2-butynylene, −CH₂C≡CCH₂− A−4.5; CA 66; 2-butyne-1,4-diyl CA 76
C_4H_6	· · ·	· · ·	· · ·	· · ·	1,2,3,4-butanetetrayl, −CH₂−CH−CH−CH₂− A−4.5; CA 66; butenylene−, −CH₂CH=CHCH₂− A−4.3; CA 66; 2-butene-1,4-diyl CA 76; butenylidene, CH₃CH=CHCH= A−4.1; CA 66
C_4H_7	· · ·	· · ·	· · ·	· · ·	butenyl, MeCH=CHCH₂− A−3.5; CA 66; butylidyne, Me(CH₂)₂C≡ A−4.1; CA 66; isobutylidyne, Me₂CHC≡ A−4.1; CA 66; 1-methylpropenylidyne, CA 76; 2-methylallyl, CH₂=CMeCH₂− A−3.5; CA 66; 2-methylpropenyl, MeMe=CH− A−3.5; A−3.6; CA 76
Cl	(mono)chlorine 1.4	chlorine(1+) ∿3.11	chloride 3.21	chloro 7.312	chloro, Cl− C−10.1; C−102; CA 76
ClF_2	(mono)chlorine difluoride 2.21; 2.22; 2.251	difluorochlorine(1+) 3.13	difluorochlorate(1−) 3.223	difluorochlorato(1−) 3.223; 7.311	difluorochloro, F₂Cl− ∿C−10.1; C−106.3
ClF_4	(mono)chlorine tetrafluoride 2.21; 2.22, 2.25	tetrafluorochlorine(1+) 3.13	tetrafluorochlorate(1−) 3.223	tetrafluorochlorato(1−) 3.223; 7.311	tetrafluorochloro, F₄Cl− ∿C−10.1; C−106.3
ClH	hydrogen chloride 2,21 2.22	hydrogenchlorine (1+) 3.13	· · ·	· · ·	chloronio, H⁺Cl− C−82.1

Atom or group	as uncharged atom, molecule or radical	as cation or cationic radical	NAME		
			an anion	as ligand	as prefix in substitutive nomenclature
ClH₂	...		chloronium, H₂Cl⁺ C-82.1	...	...	...
ClO	(mono)chlorine (mono)-oxide 2.21; 2.22	chlorosyl cation ~3.12; 3.32 oxochlorine(1+) 3.13	hypochlorite 5.214 (mono)oxochlorate(1-) 2.24	hypochlorito 7.311 (mono)oxochlorato(1-) 7.311	chlorosyl, OCl- 3.32; C-10.1; C-106.2; CA 76
ClO₂	(mono)chlorine dioxide 2.21; 2.22; 2.251	dioxochlorine(1+) 3.13 chloryl cation ~3.12; 3.32	chlorite 3.224; 5.214 dioxochlorate(1-) 2.24 3.223	chlorito 7.311 dioxochlorato(1-) 7.311	chloryl, O₂Cl- 3.32; C-10.1; C-106.2 CA 76
ClO₃	(mono)chlorine trioxide 2.21; 2.22; 2.251	trioxochlorine(1+) 3.13 perchloryl cation ~3.12; 3.32	chlorate 3.224; 5.214 trioxochlorate(1-) 2.24; 3.223	chlorato 7.311 trioxochlorato(1-) 7.311	perchloryl, O₃Cl- C-10.1; C-106.2 CA 76
ClO₄	(mono)chlorine tetraoxide 2.21; 2.22; 2.251	tetraoxochlorine(1+) 3.13	perchlorate 5.214 tetraoxochlorate(1-) 2.24; 3.223	perchlorato 7.311 CA 66 tetraoxochlorato(1-) 7.311	...
ClS SCl	(mono)sulfur monochloride 2.22	chlorosulfanyl cation 3.12 chlorosulfanilium C-83.1 chlorosulfenylium C-83.1 thiochlorosyl cation 3.12; 3.32 thiochlorine(1+) 3.11; 3.32 chlorosulfenyl cation 3.12; C-83.1	thiohypochlorite 5.211; 5.23 (mono)thiochlorate(1-) 2.24	thiohypochlorito 5.211; 5.23; 7.311 thiochlorato(1-) 7.311	chlorothio, ClS- C-621.2 chlorosulfanyl, ClS- C-81.1; C-83.1 chlorosulfeno, ClS- ~C-641.2 thiochlorosyl, SCl- 3.32
Cl₂F	dichlorine (mono)fluoride 2.21; 2.22; 2.51	fluorodichlorine(1+) 3.13	fluorodichlorate(1-) 5.211; 5.23	fluorodichlorato(1-) 5.211; 5.23; 7.311	dichlorofluoro, Cl₂F ~C-106.3
Cl₂I ICl₂	(mono)iodine dichloride 2.21; 2.22; 2.51	dichloroisodine(1+) 3.13	dichloroiodate(1-) ~3.223	dichloroiodato(1-) 7.311	dichloroiodo, Cl₂I- C-106.3; CA 66
CrO₂	(mono)chromium dioxide 2.21; 2.22; 2.251 chromyl 3.32 dioxochromium	chromyl cation 3.12 dioxochromium(2+) 3.13	...	...	...
CrO₄	...	...	chromate 5.214 tetraoxochromate(2-) 3.223	chromato 7.311 tetraoxochromato(2-) 3.223; 7.311	...

Cr₂O₇	...	...	dichromate 5.214 heptaoxodichromate(2-) 3.223	dichromato 7.311 heptaoxodichromato(2-) 7.311	...
F	(mono)fluorine 1.41	fluorine(1+) ~3.11	fluoride 7.312	fluoro, 7.312	fluoro, F- C-10.1; C-102.1; CA 76
FH	hydrogen fluoride 2.21; 2.22	hydrogenfluorine(+) 3.13	...	...	fluoronio, HF⁺- C-82.1
FH₂	...	fluoronium, H₂F⁺, 3.16; C-82.1	...	...	...
FO OF	(mono)oxygen (mono)-fluoride 2.21; 2.22	fluorosyl cation ~3.12 oxofluorine(1+) ~3.13	hypofluorite ~5.214 oxofluorate(1-)	hypofluorito 7.311 oxofluorato(1-)	fluorooxy, FO- fluorosyl, OF- ~C-10.1; C-106.2
FS SF	(mono)sulfur (mono)-fluoride 2.21; 2.22	fluorosulfur(1+) 3.13	(mono)fluorosulfate(1-) ~3.223 fluorosulfide 3.222	fluorothio ~C-621.2 fluorosulfato(1-) 3.223; 7.311	fluorothio, FS- ~C-621.2
F₂I IF₂	(mono)iodine difluor-ide 2.21; 2.22; 2.251	difluoroiodine(1+) 3.13	difluoroiodate(1-) 2.24	difluoroiodato(1-) 7.311	difluoroiodo, F₂I- ~C-10-1; C-106.3
F₂N NF₂	(mono)nitrogen difluor-ide 2.21; 2.22; 2.251	difluoronitrogen(1+) 3.13 difluoroaminyl cation ~C-83.1 difluoroaminylium ~C-83.1	difluoroamide 3.22	difluoroamido 7.311	difluoroamino, F₂N- difluoroiminio, F₂N⁺= ~C-82.2
F₂P PF₂	(mono)phosphorus difluoride 2.21; 2.22; 2.251	difluorophosphorus(1+) 3.13	difluorophosphide ~3.222 difluorophosphate(1-) 3.223	difluorophosphido difluorophosphato(1-) 7.311	difluorophosphino, F₂P-D-5.12 difluorophosphoranettriyl, F₂P≡; D-5.74 difluorophosphoranylidyne, F₂P≡ ~CA 76
F₂S SF₂	(mono)sulfur difluor-ide 2.21; 2.22; 2.251 difluorosulfane 2.3	difluorosulfur(1+) 3.14 difluorosulfoniumyl C-82.1	difluorosulfate(1-) 3.223	difluorosulfato(1-) 3.223; 7.311	difluorosulfonio, F₂S⁺- ~C-82.1
F₃I IF₃	(mono)iodine trifluor-ide 2.21; 2.22; 2.251	trifluoroiodine(1+) 3.13	trifluoroiodate(1-) 3.223	...	trifluoroiodo, -IF₃- C-10.1; C-106.3
F₃N NF₃	(mono)nitrogen tri-fluoride 2.21; 2.22; 2.251	trifluoronitrogen(1+) 3.13	trifluoronitrate(1-) 3.223	(mono)nitrogen tri-fluoride 7.321	trifluoroammonio, F₃N⁺- ~C-82.1

Atom or group	NAME				
	as uncharged atom, molecule or radical	as cation or cationic radical	an anion	as ligand	as prefix in substitutive nomenclature
F₃P PF₃	(mono)phosphorus trifluoride 2.21; 2.22; 2.251 trifluorophosphine 2.3 trifluorophosphane 2.3	trifluorophosphorus(1+) 3.13 trifluorophosphoniumyl C-83.3	trifluorophosphate(1-) 3.223	trifluorophosphine 7.321	trifluorophosphonio, F₃P⁺- ᵛC-82.1 trifluorophosphoranediyl, F₃P⟨, D-5.74 trifluorophosphoranylidene, F₃P= CA 76
F₃S SF₃	(mono)sulfur trifluoride 2.21; 2.22; 2.251	trifluorosulfur(1+) 3.14	trifluorosulfate(1-) 3.223	trifluorosulfato(1-) 3.223; 7.311	trifluorothio, F₃S- C-621.2
F₄I IF₄	(mono)iodine tetrafluoride 2.21; 2.22; 2.251	tetrafluoroiodine(1+) 3.13	tetrafluoroiodate(1-) 3.223	tetrafluoroiodato(1-) 3.223; 7.311	tetrafluoroiodo, F₄I- C-10.1 C-106.3
F₄N NF₄	. . .	tetrafluoroammonium 3.151	tetrafluoronitrate(1-) ᵛ3.223; 5.24	. . .	. . .
F₄P PF₄	(mono)phosphorus tetrafluoride 2.21; 2.22; 2.251	tetrafluorophosphonium 3.14; D-5.31 tetrafluorophosphorus(1+) 3.13	tetrafluorophosphate(1-) 3.223	tetrafluorophosphato(1-) 3.223; 7.311	tetrafluorophosphoranyl, F₄P- D-5.73
F₄S SF₄	(mono)sulfur tetrafluoride 2.21; 2.22; 2.251 tetrafluoro-λ⁴-sulfane D-0.3	tetrafluorosulfur(1+) 3.14	tetrafluorosulfate(1-) 3.223	tetrafluorosulfato(1-) 3.223; 7.311	. . .
F₅S SF₅	(mono)sulfur pentafluoride 2.21; 2.22; 2.251	pentafluorosulfur(+) 3.14	pentafluorosulfate(1-) 3.223	pentafluorosulfato(1-) 3.223; 7.311	pentafluorothio, F₅S- C-621.2
F₆I IF₆	(mono)iodine hexafluoride 2.21; 2.22; 2.251	hexafluoroiodine(1+) 3.14	hexafluoroiodate(1-) 3.223	hexafluoroiodato(1-) 7.311	hexafluoroiodo, F₆I- C-10.1; C-106.3
GeH₃	germyl D-3.43	. . .	. . .	. . .	germyl, H₃Ge- D-3.43; CA 76
H	(mono)hydrogen 1.41	hydrogen(1+) 3.11 proton	hydride 7.312	hydrido 7.312; D-2.32 hydro 7.312; 11.5 D-2.32; D-7.61	hydro, H-, A-23.1; B-1.2 C-32.1; C-161.1; CA 66
HI	hydrogen iodide 2.21; 2.22	hydrogeniodine(1+) 3.13	. . .	. . .	iodonio, HI⁺- C-82.1

20

HN NH	aminylene nitrene*	aminylene cation C-83.1	imide 3.221	imido 7.311	imino, HN= C-81.1; C-815.1; C-815.2; CA 76 epimino, -NH- (as bridge) C-815.2
HNO HON	hydroxyaminylene C-81.2	· · ·	hydroxyimide	hydroxyimido	hydroxyimino, HON= C-842.1; CA 76
HNO₂ (HO)ON	· · ·	· · ·	· · ·	· · ·	aci-nitro, (HO)ON= C-10.1; C-850.2; CA 76
HN₂	· · ·	· · ·	· · ·	· · ·	diazenyl, HN=N- CA 76 diazeno, NH=N- C-942.1; C-942.2; CA 66 1-hydrazinyl-2-ylidene, -NHN= CA 76
HN₃	hydrogen azide† 3.221; 5.1	· · ·	· · ·	· · ·	diazoamino, -N=N-NH- C-942,2; CA 66 1-triazene-1,3-diyl, -N=N-NH- CA 76
HO OH	hydroxyl 3.32	hydroxyl cation 3.32 hydrogenoxygen(1+) 3.13	hydroxide 3.221	hydroxo 7.312	hydroxy C-10.3; C-201.2; CA 76
HOP	· · ·	· · ·	· · ·	· · ·	phosphonoyl, H(O)P< D-5.66 phosphinylidene, H(O)P< CA 76
HOS	· · ·	· · ·	· · ·	· · ·	sulfeno, HOS- C-641.2; CA 76
HO₂ O₂H	perhydroxyl C-81.1 hydrogen dioxide 2.21; 2.22; 2.251	hydrogendioxygen 3.13	hydrogenperoxide 3.222	hydrogenperoxo 7.312 hydroperoxo hydroperoxy	hydroperoxy, HOO- C-218.1; CA 76
HO₂P	dioxophosphorane ∿D-5.72 oxophosphine oxide dioxo-λ⁵-phosphane D-0.3	· · ·	· · ·	· · ·	phosphinico, HO-P(O)= D-5.52; CA 76 hydrooxophosphorio D-5.69
HO₂S	· · ·	· · ·	· · ·	· · ·	sulfino, HOSO- C-641.2
HO₂Sb	· · ·	· · ·	· · ·	· · ·	stibinico, (HO)Sb(O)- ∿D-5.52; CA 76
HO₃P PHO₃	· · ·	· · ·	phosphonate(2-) 5.214; D-5.51	phosphonato 7.321	phosphonate 7.321 · · ·

* Widely used but not recommended.

† The name hydrazoic acid is not recommended 5.1.

Atom or group	NAME				
	as uncharged atom, molecule or radical	as cation or cationic radical	an anion	as ligand	as prefix in substitutive nomenclature
HO₃S HSO₃	...	...	hydrogensulfite 6.2	hydrogensulfito 6.2 sulfito(1-) 7.314	sulfo, HO-S(O)₂- C-10.3; C-641.2; CA 76
HO₃Se HSeO₃	...	...	hydrogenselenite 6.2	...	selenono, HOSe(O)₂- C-701.1; CA 76
HO₄S HSO₄	...	...	hydrogensulfate 5.214; 6.2	hydrogensulfato 6.2; 7.314	...
HO₆P₂	...	...	diphosphate(III,V), O₂PH-O-PO₃³⁻ 4.13; 5.214	...	...
HP	...	...	hydrogenphosphide(2-) 3.22	hydrogenphosphido 3.221; 7.311	phosphinediyl, HP= D-5.14 phosphinidene, HP= C-81.1; CA 76 phosphoranetetrayl, HP≡ ∿D-5.74
HPS	thioxophosphine	...	...	...	thiophosphonoyl, D-5.67 phosphonothioyl, D-5.67 phosphinothioylidene, HP- (=S)= CA 76
HS	sulfhydryl	hydrogensulfur(1+) 3.13 sulfanylium C-83.1	hydrogensulfide(1-) 6.2 hydrosulfide	mercapto 7.312 hydrosulfido 7.311	mercapto C-10.3; C-511.1; CA 76 sulfoniumylidene, HS⁺= ∿C-83.3
HSb	...	...	hydrogenantimonide 3.222	hydrogenanthimonido 7.311	stibinediyl, HSb= D-5.14 stibylene, HSb= CA 76
HSe	(mono)hydrogen (mono)selenide 2.21; 2.22; 2.251	selanylium ∿C-83.1; C-701 hydrogenselenium(1+) 3.13	hydrogenselenide 3.221; ∿6.2 hydroselenide	hydrogenselenido 7.311	hydroseleno, HSe- C-701.1 selenyl HSe- CA 76
HSi	...	...	...	silanetriyl D-6.74	silanetriyl, HSi≡ ∿D-6.12 silylidyne, HSi≡ CA 76
HTe	(mono)hydrogen (mono)- telluride 2.21; 2.22	hydrogentellurium(1+) 3.13 tellanylium C-83.1 C-701	hydrogentelluride 3.221; 6.2 hydrotelluride	hydrogentellurido 7.312 hydrotellurido	hydrotelluro, HTe- C-701.1 telluryl, HTe- CA 76

Formula					
H_2I	. . .	iodonium H_2I^+ C-82.1	. . .	. . .	. . .
H_2IO_2	. . .	. . .	. . .	. . .	dihydroxyiodo, $(HO)_2I-$ C-10.1; C-106.3; CA 66; see CA 76 §188
$H_2N\ NH_2$	<u>aminyl</u> C-81.2	<u>aminyl</u> cation C-83.1 aminylium C-83.1	<u>amide</u> 3.221	<u>amido</u> 7.311; CA 66	amino, H_2N- C-811.1; C-812.2; CA 76 iminio, $H_2N^+=$ ~C-81.1; C-82.2
H_2NO	aminooxyl	oxoammonium	hydroxylamide 3.221	hydroxylamido-$\underline{O}$ 7.311 hydroxylamido-$\underline{N}$ 7.33	aminooxy, H_2NO- CA 66 hydroxyamino, HONH- C-841.1; CA 76
H_2N_2	diazene D-4.12	. . .	. . .	. . .	hydrazono, $H_2N-N=$ C-921.1; CA 76 hydrazo, -NHNH- C-921.1; CA 76* hydrazi, -NHNH- C-921.4; CA 76†
H_2N_3	triazenyl D-4.14	. . .	. . .	. . .	triazenyl, $H_2N-N=N-$ CA 76 triazeno C-942.3; CA 66
H_2N_4	. . .	. . .	. . .	. . .	1,1,4,4-tetrazanetetrayl, $=N(NH)_2N=$ D-4.14 1,4-tetrazanediylidene, $=N(NH)_2N=$ CA 76
H_2O	<u>water</u> dihydrogen oxide 2.21; 2.22; 2.251	dihydrogenoxygen(1+) C-82.1 oxoniumyl C-82.1	. . .	<u>aqua</u>‡ 7.322	oxonio, H_2O^+- C-82.1; C-85; C-87.1 oxoniumyl, H_2O^+- ~C-83.3
H_2OP $H_2P(O)-$	. . .	. . .	. . .	. . .	phosphinoyl, $H_2P(O)-$ D-5.66 phosphinyl, $H_2P(O)-$ CA 76
H_2O_2	(di)hydrogen dioxide (di)hydrogen peroxide 2.21; 2.22; 2.251	. . .	. . .	. . .	. . .
H_2O_2P PH_2O_2	. . .	. . .	phosphinate, H_2PO_2- 5.214; D-5.51	phosphinato 7.321	hydroxyphosphinyl, $(HO)_2P(O)O-$ CA 76
H_2O_3P $HPHO_3$	. . .	. . .	hydrogenphosphonate $HPO_2(OH)-$ 5.214; 6.2; D-5.51	hydrogenphosphonato 7.321	phosphono, $(HO)HP(O)-$ D-5.52; CA 76

* to different atoms † to a single atom ‡ aquo in older literature

23

Atom or group	NAME				
	as uncharged atom, molecule or radical	as cation or cationic radical	an anion	as ligand	as prefix in substitutive nomenclature
H_2O_4P H_2PO_4	...	...	dihydrogenphosphate, $PO_2(OH)_2^-$ 5.214; 6.2	dihydrogenphosphato 7.321	phosphonooxy, $(HO)_2P(O)O-$
H_2O_5P $P_2H_2O_5$	...	...	diphosphonate (IV), $O_2PH-PHO_2^{2-}$ 5.214	...	...
H_2P	dihydrogen phosphorus	dihydrogenphosphorus-(1+)	dihydrogenphosphphide(1-) 3.221	dihydrogenphosphido 3.221; 7.311	phosphino, H_2P- C-81.1; D-5.12; CA 76 phosphoranetriyl, $H_2P\leqq$ D-5.74 phosphoranylidyne CA 76
H_2PS	...	...	...	...	thiophosphinoyl, $H_2P(=S)-$ D-5.67 phosphinothioyl, $H_2P(=S)-$ D-5.67; CA 76
H_2P_2	...	...	...	...	1,2-diphosphanediyl, $-HP-PH-$ D-4.14 1,2-diphosphinediyl, $-HP-PH-$ CA 76
H_2S	(di)hydrogen (mono)-sulfide 2.21; 2.22; 2.251; sulfane 2.3	dihydrogensulfur(1+) 3.13; sulfaniumyl C-83.3	dihydrosulfate(1-) 3.223	hydrogen sulfide 7.321	sulfonio, H_2S^+- C-82.1; C-85; C-87.1 sulfoniumyl, H_2S^+- ∿C-83.3
H_2Sb	...	...	dihydrogenantimonide 3.223	dihydrogenantimonido 7.311	stibino, H_2Sb-* D-5.12; CA 76
H_2Se	selane 2.3 (di)hydrogen (mono)-selenide 2.21; 2.22; 2.251	dihydrogenselenium(1+) 3.13; selenoniumyl C-83.3	dihydroselenate(1-) 3.223	hydrogen selenide 7.321	selenonio, H_2Se^+- C-82.1; C-85; C-87.1 selenoniumyl ∿C-83.3
H_2Si	...	...	...	silanediyl D-6.74 silylene	silanediyl, $H_2Si=$ D-6.12; D-6.74 silylene, $H_2Si=$ C-81.1; CA 76
H_2Te	(di)hydrogen telluride 2.21; 2.22; 2.251	dihydrogentellurium(1+) 3.13	...	hydrogen telluride	...
H_3N	ammonia 2.3	ammoniumyl C-83.3	...	ammine 7.322	ammonio, H_3N^+- C-82.1; C-85; C-816.3; C-87; C-816.3 ammoniumyl, H_3N^+- C-83.3; C-86.1

* Stibyl has been used for H_2Sb- but is not recommended.

Formula	Name	Cation	Anion		Substituent group
H₃NO H₂NOH	hydroxylamine 3.152 C-8.4	· · ·	· · ·	· · ·	· · ·
H₃NP	· · ·	· · ·	· · ·	· · ·	phosphonimidoyl, H₂P(=NH)- D-5.67 phosphinimyl, H₂P(=NH)- CA 76
H₃NS H₂NSH	thiohydroxylamine	· · ·	· · ·	· · ·	· · ·
H₃N₂	hydrazyl C-82.1	hydrazyl cation C-83.1 hydrazylium C-83.1	hydrazide 3.221	hydrazido 7.311	hydrazino, NH₂NH- C-921.1; CA 76
H₃O	trihydrogen oxide 2.21; 2.22; 2.251	oxonium 3.14 trihydrogenoxygen(1+) 3.13	· · ·	· · ·	· · ·
H₃OP H₃PO	phosphine oxide D-5.41 oxopyosphorane D-5.41 oxotrihydrophosphorus D-5.41	· · ·	· · ·	· · ·	· · ·
H₃OSi H₃SiO	· · ·	· · ·	silanolate D-6.85	silanolato siloxy, H₃SiO- D-6.79	siloxy, H₃SiO- D-6.791 CA 76
H₃P PH₃	phosphine 2.3; D-5.11 phosphane 2.3	phosphoniumyl C-83.3	· · ·	phosphine 7.321	phosphonio, H₃P⁺- C-83.1; D-5.33 phosphoniumyl, H₃P⁺- C-83.3 D-5.32 phosphoranediyl, H₃P= D-5.74 phosphoranylidene, H₃P= CA 76
H₃P₂	· · ·	· · ·	· · ·	· · ·	dipyosphanyl, H₂P-PH- D-4.14 diphosphino, H₂P-PH- CA 66 diphosphinyl, CA 76
H₃P PbH₃	· · ·	plumbylium C-81.1; C-831; D-3.43	· · ·	· · ·	plumbyl, H₃P- C-81.1; D-3.43; CA 76
H₃S SH₃	trihydrogen (mono)- sulfide 2.21; 2.22; 2.251	sulfonium 3.14; C-82.1 sulfanium C-82.2 trihydrogensulfur(1+) 3.13	trihydrosulfate(1-) 3.232	· · ·	· · ·
H₃Sb SbH₃	stibine 2.2; D-5.11 D-3.43 stibane 2.3	trihydrogenantimony	· · ·	stibine 7.321 stibane 7.321	stibonio, H₃Sb⁺- ~C-82.1; D-5.34 stiboniumyl, H₃Sb C-83.3

Atom or group	NAME				
	as uncharged atom, molecule or radical	as cation or cationic radical	an anion	as ligand	as prefix in substitutive nomenclature
H_3Se SeH_3	trihydrogen(mono)-selenide 2.21; 2.22; 2.251	selanium 2.3; C-87.2; selenonium C-82.1; trihydrogenselenium(1+) 3.13	trihydroselenate(1-)	···	···
H_3Si	silyl C-81.1; D-6.12	silylium C-81; C-81.1; C-83.1; D-6.12	···	silyl D-6.74	silyl, H_3Si- D-6.12; CA 76
H_3Sn	···	···	···	···	stannyl, H_3Sn D-3.43; CA 76
H_3Te	trihydrogen telluride	telluronium, H_3Te^+ 3.14; C-82.1	trihydrotellurate(1-)	···	···
H_4N NH_4	···	ammonium 3.14; C-82.1	···	···	···
H_4N_2 N_2H_4	hydrazine 2.3	hydraziniumyl C-83.3	···	hydrazine 7.321	hydrazinio, H_3N^+ -NH C-82.1; C-82.2; hydraziniumyl, H_3N^+ -NH- C-83.3
H_4N_3 N_3H_4	···	···	···	···	triazano, $H_2NNHNH-$ C-942.3; CA 66; triazanyl, $H_2NNHNH-$ CA 76
H_4P PH_4	···	phosphonium 3.14; D-5.31	···	···	phosphoranyl, H_4P- D-5.74; CA 76
H_4P_2 P_2H_4	diphosphane 2.3; D-4.1	···	···	···	···
H_4Sb	···	stibonium ᐱC-82.1	···	···	···
H_4Si SiH_4	silane 2.3; D-6.11	···	···	silane 7.321	···
H_5N_2 N_2H_5	···	hydrazinium(1+) 3.17; hydrazinium, C-82.2; C-921.6	···	hydrazinium(1+) 7.324	···
H_5N_4	···	···	···	···	tetrazanyl, $H_2N(NH)_2NH-$ D-4.14
H_5OSi_2	disiloxanyl C-81.1; D-6.51; disilanoxyl C-81.1; D-6.79	···	disilanolate C-206.1; D-6.82; disilanyl oxide C-206	disilanolato 7.314	disilanyloxy, H_3SiSiH_2O- D-6.79; CA 76; disilanoxy, H_3SiSiH_2O- CA 66

			· · ·	· · ·	disiloxanyl, $H_3SiOSiH_2^-$ D-6.79
H_5P PH_5	phosphorane D-5.71; λ^5-phosphane D-0.3	· · ·	· · ·	· · ·	· · ·
H_5Si_2	disilanyl C-81.1; D-6.12	· · ·	· · ·	· · ·	disilanyl, $H_3SiSiH_2^-$ D-6.12 CA 76
H_6NSi_2	disilazanyl C-81.1; D-6.51; disilanylaminyl C-81.1; D-6.5	· · ·	disilazanate D-6.87; disilylamide 7.311; disilanylamide 7.311	disilazanato, $(H_3Si)_2N^-$ D-6.87; disilylamido 7.311; disilanylamido 7.311	disilylamino, $(H_3Si)_2N^-$ D-6.79; CA 76; 2-disilazanyl D-6.51; disilazanyl $H_3SiNHSiH_2^-$ D-6.79; disilanylamino, $H_3SiSiH_2NH^-$ D-6.79
H_6N_2 N_2H_6	· · ·	hydrazinium(2+) C-921.6; 3.17			
I	(mono)iodine 1.4	iodine(1+) 3.11	iodide	iodo 7.312	iodo, I^- C-10.1; C-102.1; CA 76
IO	(mono)iodine (mono)-oxide 2.21; 2.22; 2.251	iodosyl cation 3.12; oxoiodine(1+) 3.13	hypoiodite 5.211; (mono)oxoiodate(1-) 2.24	hypoiodito 7.311; (mono)oxoiodato(I) 7.311	iodosyl, OI^- C-10.1; C-106.1; CA 76; iodoso, OI^- CA 66
IO_2	· · ·	iodyl	iodite 3.224; 5.211; dioxoiodate(1-) 2.24	iodito 7.311; dioxoiodato(III) 7.311	iodyl, O_2I^- C-10.1; C-106.1; CA 76; iodoxy, O_2I^- CA 66
IO_3	· · ·	· · ·	iodate; trioxoiodate(1-) 2.24	iodato; trioxoiodato(1-) 7.311	· · ·
IO_4	· · ·	· · ·	periodate; tetraoxoiodate(1-)	periodato; tetraoxoiodato(1-) 7.311	· · ·
I_3	triiodide 1.4	triiodide(1+) 1.4; 3.11	triiodide 3.221	triiodo 7.31	· · ·
MnO_4	· · ·	· · ·	permanganate, MnO_4^-; tetraoxomanganate, MnO_4^- 5.214; manganate, MnO_4^- 5.214; tetraoxomanganate(2-) 5.214	permanganato 7.311; tetraoxomanganato 7.311; manganato 7.311; tetraoxomanganato(2-) 7.311	· · ·
N	(mono)nitrogen 1.4	nitrogen(1+) ~3.11	nitride 3.21	nitrido 7.311	nitrilo, $N\equiv$ C-72.1; C-81.1; C-815.1; CA 76

	NAME				
Atom or group	as uncharged atom, molecule or radical	as cation or cationic radical	an anion	as ligand	as prefix in substitutive nomenclature
NO	(mono)nitrogen (mono)oxide 2.21; 2.22	nitrosyl cation 3.12 oxoaminylium C-81.2	oxoamide 3.22 (mono)oxonitrate(1-) 5.214	nitrosyl 7.323	nitroso, O=N- C-10.1; C-851.1; CA 76
NO_2	(mono)nitrogen dioxide 2.21; 2.22; 2.251	nitryl cation 3.12	nitrite, NO_2^- 5.214 dioxonitrate(1-) NO_2^- 2.24 nitroxylate, NO_2^{2-}	nitro 7.33 nitrito-N ~7.33 nitrito-O ~7.33 dioxonitrate(1-) 2.24 nitroxylato 5.214; 7.311	nitro, O_2N- C-10.1; CA 76 nitrosooxy, O=N-O-
NO_3	(mono)nitrogen trioxide 2.21; 2.22; 2.251	· · ·	nitrate 5.214	nitrato 7.311	nitrooxy, O_2N-O-
NO_4	· · ·	· · ·	peroxonitrate 5.214		
NS	(mono)nitrogen (mono)sulfide 2.21; 2.22	thionitrosyl cation 3,12; 2.32 thioxoaminylium C-81.2	thioxoamide 3.22	thionitrosyl 3.12; 3.32	thionitroso, S=N C-10.1; C-502; CA 76
N_2	dinitrogen 1.4	dinitrogen(1+), N_2^+ ~3.11	· · ·	dinitrogen 7.321; CA 76	azo, -N=N- C-911; C-912; CA 76 azino, =N-N= C-923.1; CA 76 azi, -N=N-* C-931.5; CA 76 diazo, $N_2=$ C-10.1; C-931.4; CA 76 diazonio, N_2^+- CA 76; C-931.1
N_2O	dinitrogen (mono)oxide 2.21; 2.22; 2.251	· · ·	· · ·	dinitrogen (mono)oxide	azoxy, -N(O)=N- C-913.1; CA 76 nitrosoimino, ON-N= CA 76 nitrosimino, ON-N= CA 66
N_2O_2	dinitrogen dioxide 2.21; 2.22; 2.251	· · ·	hyponitrite(2-) $\overline{O}NN\overline{O}$ 5.214	hyponitrito, ONNO 7.311	nitroimino, $O_2N-N=$ CA 76
N_3	trinitrogen 1.4	trinitrogen(1+) ~3.11	azide 3.221	azido 7.311	azido, N_3^- C-10.1; C-941.1; CA 76 diazolimino, -N=N-N= C-942.2 triazirinyl, $N=N-\bar{N}-$ B-1.1; B-1.3; B-5.11

* both valencies attached to the same atom

28

	(mono)oxygen 1.4	oxygen(1+) ~3.11	oxide 3.21	oxo 7.312	
o	(mono)oxygen 1.4	oxygen(1+) ~3.11	oxide 3.21	oxo 7.312	oxo, O= C-10.3; C-316; CA 76; oxy, -O- C-72.2; C-212.1; CA 76; epoxy, -O- (as bridge) C-212.2; CA 76; oxido, -O⁻ C-86.2
OP PO	(mono)phosphorus (mono)oxide 1.4	phosphoryl cation 3.12; 3.32; oxophosphorus(1+)	. . .	. . .	oxophosphorio- OP- D-5.69; phosphoroso, OP- CA 76; phosphoryl, OP≡ 3.32; D-5.66; phosphinylidyne, OP≡ CA 76
OS SO	(mono)sulfur, (mono)- oxide 2.21; 2.22; oxosulfane 2.3	sulfinyl* cation 3.32; oxosulfoniumyl; oxosulfur(2+) 3.13	. . .	sulfur monoxide 7.321	sulfinyl*, OS= C-631.2; CA 76
OSe SeO	(mono)selenium (mono)oxide 2.21; 2.22; oxoselane 2.3	seleninyl cation 3.32; oxoselenium(1+) 3.14; oxoselenoniumyl C-83.3	selenoxide; oxoselenate(2-) 3.223	selenium monooxide 7.321	seleninyl, OSe=, -Se(O)- C-701.1; CA 76
OTi	. . .	(mono)oxotitanium(2+)† TiO²⁺	. . .	. . .	. . .
OV	. . .	(mono)oxovanadium(2+)† VO²⁺	. . .	. . .	. . .
OZr	. . .	(mono)oxozirconium(2+)† ZrO²⁺	. . .	. . .	. . .
O₂	dioxygen 1.4	dioxygen(1+) ~3.11; dioxygen(2+) ~3.11; dioxygenyl(1+)**	peroxide, O₂²⁻ 3.221; hyperoxide, O₂⁻†	peroxo, O₂²⁻ 7.312; dioxygen, O₂⁻; hyperoxo O₂⁻‡	dioxy, -OO- C-218.2; CA 76; epidioxy, -OO- (as bridge) C-218.2; CA 76
O₂P	(mono)phosphorus dioxide 2.21; 2.22; 2.251	. . .	dioxophosphate(1-) 5.214	. . .	phosphinato, -O₂P- D-5.52; phospho, O₂P- CA 76
O₂P₂S P₂O₂S	diphosphorus dioxide sulfide 2.21; 2.22; 2.251	thiodiphosphoryl; thiopyrophosphoryl	. . .	. . .	. . .
O₂S SO₂	(mono)sulfur dioxide 2.21; 2.22; 2.251	sulfonyl# cation 3.32; dioxosulfur(2+) 3.13	sulfoxylate 5.214	sulfur dioxide; sulfoxylato 5.214; 7.311	sulfonyl#, O₂S=, -S(O)₂- C-631.1; C-631.2; CA 76; sulfinato, -O₂S- ~C-641.2

* The name thionyl is no longer used.   † The names titanyl, vanadyl and zirconyl are not recommended.
‡ Although superoxide and superoxido have been used they are not recommended. In some languages "super" is equivalent to "per".
** Not recommended.   # The name sulfuryl is no longer used.

Atom or group	as uncharged atom, molecule or radical	as cation or cationic radical	an anion	as ligand	as prefix in substitutive nomenclature
			NAME		
O_2S_2 S_2O_2	...	...	thiosulfite 3.224; 5.23	...	thiosulfonato,$-O_2S_2-$ C-86.1; C-641.6
O_2Se SeO_2	(mono)selenium dioxide 2.21; 2.22; 2.251	selenonyl cation 3.32 dioxoselenium(1+) 3.13	selenone selenoxylate(2-) ~5.214; 5.23	selenium dioxide 7.321 selenoxylato(2-) 5.214; 7.311	selenonyl, $O_2Se=$, $-Se(O)_2-$ C-701.1; CA 76
O_2U UO_2	...	uranyl(2+) 3.32 dioxouranium(2+) 3.13	...	...	...
O_3	trioxygen (ozone) 1.4	trioxygen(1+) 1.4; 3.11	ozonide, O_3^- 3.221	ozonido 7.311	trioxy, $-O-O-O-$ ~C-515.1; C-515.4 epitrioxy, $-OOO-$ (as bridge) ~C-218.2
O_3P PO_3	...	diphosphoryl pyrophosphoryl	phosphite, PO_3^{3-}* metaphosphate, $(PO_3^-)_n$ 5.214	phosphito 7.321	...
O_3P_2 P_2O_3	tetraphosphorus hexa-oxide, P_4O_6 2.21; 2.22; 2.251 phosphorus(III) oxide	...	...	...	...
O_3Re	rhenium trioxide	...	trioxorhenate(1-) 5.214	...	...
O_3S SO_3	(mono)sulfur trioxide 2.21; 2.22; 2.251 trioxo-λ^6-sulfane D-0.3	trioxosulfur(1+) 3.14	sulfite 5.214	sulfito 5.214; 7.311 peroxosulfoxylato	sulfonato, $-O_3S-$ C-86.1
O_3S_2 S_2O_3	disulfur trioxide 2.21; 2.22; 2.251	trioxodisulfur(1+) 3.13	thiosulfate 5.214; 5.23 thioperoxosulfoxylate 5.214; 5.22; 5.23	(mono)thiosulfato 7.311 thiodiperoxosulfoxylato disulfoxylato	...
O_3Se SeO_3	...	...	silenite 5.24 trioxoselenate(2-) 5.214	selenito 7.311 trioxoselenato(2-) 7.311	...
O_3Si SiO_3	...	...	metasilicate 5.213	...	...

* known only as esters 3.224.

Formula	Oxide / parent name	Cation	Anion (-ate)	Acyl (-ato)	Index notes
O_4P PO_4			phosphate 5.214	phosphato 7.311	· · ·
O_4Re ReO_4			perrhenate, ReO_4^- 5.214 tetraoxorhenate(1-) rhenate, $ReO_4^=$ 5.214 tetraoxorhenate(2-) 5.214	perrhenato, ReO_4^- 7.311 tetraoxorhenate(1-) ReO_4^- rhenate, $ReO_4^=$ 7.311 tetraoxorhenate(2-) ReO_4^{2-} 7.311	· · ·
O_4S SO_4		tetraoxosulfur (1+) 3.13	sulfate 5.214	sulfato 7.311	sulfonylbis(oxy) -$OS(O)_2$-O- CA 76 sulfonyldioxy, -$OS(O)_2$-O- C-72.2; C-205.2; CA 66
O_4S_2 S_2O_4	(mono)sulfur tetra-oxide 2.21; 2.22; 2.251		dithionite 3.224; 5.214	dithionito 7.311	· · ·
O_4Se SeO_4			selenate 5.214	selenato 7.311	· · ·
O_4Si SiO_4			orthosilicate 5.214	· · ·	· · ·
O_4Tc TcO_4			pertechnetate, TcO_4^- 5.214 technetate, TcO_4^{2-} 5.214	· · ·	· · ·
O_5P PO_5			peroxomonophosphate, PO_5^{3-} 5.214	· · ·	· · ·
O_5S SO_5			peroxomonosulfate 5.214; 5.22	· · ·	· · ·
O_5S_2 S_2O_5	disulfuryl 3.32 disulfur pentaoxide 2.21; 2.22; 2.251	pentaoxodisulfur(1+) 3.13	disulfite 5.214	disulfito 5.214; 7.311	· · ·
O_6P_2 P_2O_6			hypophosphate, O_3P-PO_3^{4-} diphosphate(IV), O_3P-PO_3^{4-} 5.214	· · ·	· · ·
O_6S_2 S_2O_6			dithionate 5.214	dithionato 5.214; 7.311	· · ·
O_6Te TeO_6			orthotellurate 5.214	orthotellurato 7.311	· · ·
O_7P_2 P_2O_7	diphosphorus hepta-oxide 2.21; 2.22; 2,251		diphosphate 5.214 pyrophosphate 5.214	diphosphato 7.311 pyrophosphato 7.311	· · ·

31

Atom or group	NAME				
	as uncharged atom, molecule or radical	as cation or cationic radical	an anion	as ligand	as prefix in substitutive nomenclature
O_7S_2 S_2O_7	···	···	disulfate 4.12; 5.214	disulfato 5.214; 7.311	···
O_7Si_2 Si_2O_7	···	···	disilicate, $O_3Si-O-SiO_3$ 4.12	···	···
O_8P_2 P_2O_8	···	···	peroxodiphosphate, $O_3P-O-PO_3^{4-}$ 5.214	···	···
O_8S_2 S_2O_8	···	···	peroxodisulfate 5.214	peroxodisulfato 5.214 7.311	···
P	(mono)phosphorus 1.4	phosphorus(1+) ~3.11	phosphide 3.21	phosphido 7.311	phosphinetriyl, P≡ D-5.15 phosphinidyne, P≡ CA 76
P_2	diphosphorus 1.4	···	···	···	diphosphanetetrayl, =P-P= D-4.14 1,2-diphosphinediylidene, =P-P= CA 76 diphosphinetetrayl, =P-P= CA 66 phosphoro, -P=P- CA 66 1,2-diphosphenediyl CA 76
S	(mono)sulfur 1.4	sulfur(1+) ~3.11	sulfide 3.21	thio 7.312	thioxo, S= C-532.3; CA 76 thio, -S- C-72.1; C-514.1; C-661.3; CA 76 epithio, -S- (as bridge) C-514.4 sulfido -S- C-86.2; C-511.4
S_2	disulfur 1.4	disulfur(1+) ~3.11	disulfide 3.221	disulfido 7.312	dithio, -SS- C-515.1; C-515.3; CA 76 epidithio, -SS- (as bridge) C-515.4; CA 76 thiosulfonyl, S(S)= C-641.6 perthio CA 76
S_4	tetrasulfur 1.4	tetrasulfur(1+) ~3.11 tetrasulfur(2+) ~3.11	tetrasulfide ~3.211 trithioperoxosuloxylate 5.22; 5.23 trithiosulfite 3.224 5.23	tetrasulfido	tetrathio, -SSSS- CA 76 epitetrathio, -SSSS- (as bridge) C-515.4 tetrasulfanediyl, -SSSS- ~C-515.3

Sb	(mono)antimony 1.4	antimony(1+) ~3.11	antimonide ~2.21; 3.21	antimonido 7.311	stibinetriyl, Sb≡ D-5.15; stibylidyne, Sb≡ CA 76
Sb₂	diantimony 1.4	diantimony(1+) ~3.11	. . .	. . .	antimono, -Sb=Sb-* CA 66
Se	(mono)selenium 1.4	selenium(1+) ~3.11	selenide 3.21	seleno ~7.312	seleno, -Se- C-701.1; CA 76; episeleno, -Se- (as bridge) C-701; CA 76; selenoxo, Se= C-701.1; CA 76
Se₂	diselenium 1.4	deselenium(2+) ~3.11	diselenide ~3.221	diselenido ~7.312	diseleno, -SeSe- C-701.1; epidiseleno, -SeSe- (as bridge) CA 76; perseleno, Se=Se= CA 76
Se₄	tetraselenium 1.4	tetraselenium(2+) 3.11	tetraselenide 3.221	. . .	tetraseleno, -SeSeSeSe- C-515.1; C-701.1 epitetraseleno (as bridge) C-515.4; C-701.1
Si	(mono)silicon 1.4	. . .	silicide 3.21	silicido 7.311	silanetetrayl, =Si= ~D-6.1; CA 76
Si₂	disilicon 1.4	. . .	. . .	. . .	disilanehexayl, ≡Si-Si≡ D-6.12; disilanediylidyne, ≡Si-Si≡ CA 66
Te	(mono)tellurium 1.4	tellurium(1+) 3.11	telluride 3.21	telluro ~7.312	telluro, -Te- C-701.1; CA 76 epitelluro, -Te- (as bridge) C-701.1; telluroxo, Te= C-701.1; CA 76
Te₄	tetratellurium 1.4	tetratellurium(2+) ~3.11; 3.14	tetratelluride ~7.312	tetratellurido ~7.312	tetratelluro, -TeTeTeTe- C-515.1; C-701.1 epitetratelluro, -TeTeTeTe- (as bridge) C-515.1; C-701.1

33

* The names antimono and 1,2-distibinediyl have been used for -Sb=Sb- and distibenyl for HSb=Sb- (cf. CA 76). Inasmuch as the substances once thought to contain these groups are known to be polymeric, there is no need for these names except, possibly, as class names.

TABLE OF ATOMIC WEIGHTS 1975

(Scaled to the relative atomic mass, $A_r(^{12}C) = 12$)

The atomic weights of many elements are not invariant but depend on the origin and treatment of the material. The footnotes to this Table elaborate the types of variation to be expected for individual elements. The values of $A_r(E)$ given here apply to elements as they exist naturally on earth and to certain artificial elements. When used with due regard to the footnotes they are considered reliable to ±1 in the last digit or ±3 when followed by an asterisk*. Values in parentheses are used for certain radioactive elements whose atomic weights cannot be quoted precisely without knowledge of origin; the value given is the atomic mass number of the isotope of that element of longest known half life.

| | | Alphabetical order in English | | |
Name	Symbol	Atomic number	Atomic weight	Footnotes
Actinium	Ac	89	227.0278	z
Aluminium	Al	13	26.98154	
Americium	Am	95	(243)	
Antimony	Sb	51	121.75*	
Argon	Ar	18	39.948*	w, x
Arsenic	As	33	74.9216	
Astatine	At	85	(210)	
Barium	Ba	56	137.33	x
Berkelium	Bk	97	(247)	
Beryllium	Be	4	9.01218	
Bismuth	Bi	83	208.9804	
Boron	B	5	10.81	w, y
Bromine	Br	35	79.904	
Cadmium	Cd	48	112.41	x
Caesium	Cs	55	132.9054	
Calcium	Ca	20	40.08	x
Californium	Cf	98	(251)	
Carbon	C	6	12.011	w
Cerium	Ce	58	140.12	x
Chlorine	Cl	17	35.453	
Chromium	Cr	24	51.996	
Cobalt	Co	27	58.9332	
Copper	Cu	29	63.546*	w
Curium	Cm	96	(247)	
Dysprosium	Dy	66	162.50*	
Einsteinium	Es	99	(254)	
Erbium	Er	68	167.26*	
Europium	Eu	63	151.96	x
Fermium	Fm	100	(257)	
Fluorine	F	9	18.998403	
Francium	Fr	87	(223)	
Gadolinium	Gd	64	157.25*	x
Gallium	Ga	31	69.72	
Germanium	Ge	32	72.59*	
Gold	Au	79	196.9665	
Hafnium	Hf	72	178.49*	
Helium	He	2	4.00260	x
Holmium	Ho	67	164.9304	
Hydrogen	H	1	1.0079	w
Indium	In	49	114.82	x
Iodine	I	53	126.9045	
Iridium	Ir	77	192.22*	
Iron	Fe	26	55.847*	
Krypton	Kr	36	83.80	x, y
Lanthanum	La	57	138.9055*	x
Lawrencium	Lr	103	(260)	
Lead	Pb	82	207.2	w, x
Lithium	Li	3	6.941*	w, x, y
Lutetium	Lu	71	174.97	
Magnesium	Mg	12	24.305	x
Manganese	Mn	25	54.9380	
Mendelevium	Md	101	(258)	
Mercury	Hg	80	200.59*	

Name	Symbol	Alphabetical order in English Atomic number	Atomic weight	Footnotes
Molybdenum	Mo	42	95.94	
Neodymium	Nd	60	144.24*	x
Neon	Ne	10	20.179*	y
Neptunium	Np	93	237.0482	z
Nickel	Ni	28	58.70	
Niobium	Nb	41	92.9064	
Nitrogen	N	7	14.0067	
Nobelium	No	102	(259)	
Osmium	Os	76	190.2	x
Oxygen	O	8	15.9994*	w
Palladium	Pd	46	106.4	x
Phosphorus	P	15	30.97376	
Platinum	Pt	78	195.09*	
Plutonium	Pu	94	(244)	
Polonium	Po	84	(209)	
Potassium	K	19	39.0983*	
Praseodymium	Pr	59	140.9077	
Promethium	Pm	61	(145)	
Protactinium	Pa	91	231.0359	z
Radium	Ra	88	226.0254	x, z
Radon	Rn	86	(222)	
Rhenium	Re	75	186.207	
Rhodium	Rh	45	102.9055	
Rubidium	Rb	37	85.4678*	x
Ruthenium	Ru	44	101.07*	x
Samarium	Sm	62	150.4	x
Scandium	Sc	21	44.9559	
Selenium	Se	34	78.96*	
Silicon	Si	14	28.0855*	
Silver	Ag	47	107.868	x
Sodium	Na	11	22.98977	
Strontium	Sr	38	87.62	x
Sulfur	S	16	32.06	w
Tantalum	Ta	73	180.9479*	
Technetium	Tc	43	(97)	
Tellurium	Te	52	127.60*	x
Terbium	Tb	65	158.9254	
Thallium	Tl	81	204.37*	
Thorium	Th	90	232.0381	x, z
Thulium	Tm	69	168.9342	
Tin	Sn	50	118.69*	
Titanium	Ti	22	47.90*	
Tungsten (Wolfram)	W	74	183.85*	
Uranium	U	92	238.029	x, y
Vanadium	V	23	50.9414*	
Xenon	Xe	54	131.30	x, y
Ytterbium	Yb	70	173.04*	
Yttrium	Y	39	88.9059	
Zinc	Zn	30	65.38	
Zirconium	Zr	40	91.22	x

w Element for which known variations in isotopic composition in normal terrestrial material prevent a more precise atomic weight being given; $A_r(E)$ values should be applicable to any "normal" material.

x Element for which geological specimens are known in which the element has an anomalous isotopic composition, such that the difference in atomic weight of the element in such specimens from that given in the Table may exceed considerably the implied uncertainty.

y Element for which substantial variations in A_r from the value given can occur in commercially available material because of inadvertent or undisclosed change of isotopic composition.

z Element for which the value of A_r is that of the radioisotope of longest half-life.

0 0 8 021982 9 (Flexicove

DATE DUE			
Chemistry Dept			